GENE THERAPY

AND
HUMAN GENETIC ENGINEERING

by
Helen Watt

*All booklets are published thanks to the
generous support of the members of the
Catholic Truth Society*

CATHOLIC TRUTH SOCIETY
PUBLISHERS TO THE HOLY SEE

2

CONTENTS

THE LINACRE CENTRE

The Linacre Centre is the only Catholic institution of its kind specialising in the field of healthcare ethics in Great Britain and Ireland. As such it provides a unique service to the Catholic community in these islands and more particularly to Catholics working in the field of healthcare. The Centre also exists to assist the teaching authorities within the Church in addressing bioethical issues, and to communicate and defend the Church's moral teaching in debates over public policy and legislation in the United Kingdom.

The Centre has built up a large bioethics library at the Hospital of St John and St Elizabeth in London. It has three fulltime research fellows who are able to give time and thought to new and difficult issues in bioethics and it is also able to call upon the help of a range of experts in medicine, law, philosophy, theology and history. The Centre is affiliated to the Ave Maria School of Law, Ann Arbor, Michigan. It publishes reports, organises conferences and lectures, and does consultancy work for individuals and for other organizations. The cooperation of the Linacre Centre with the CTS Explanations series of booklets is intended to advance this work of providing clear Catholic teaching on bioethical issues.

Introduction

Human genetics is a new science, which is very rapidly advancing. Scientists are finding out more and more about the location, structure and function of genes, in what is called the *Human Genome Project*. We now have a 'rough draft' of the human *genome* (the pattern of genes within each cell of our bodies), although the function of many genes is still unclear. There is much excitement at the possibilities offered by the knowledge of genes we are acquiring, although there are also fears that this knowledge will not be well-used.

Genetic knowledge is good in itself, like other kinds of scientific knowledge. The practical uses to which it is put, and for which it is acquired, can be either good or bad. Unfortunately, the current situation is that we can diagnose many more conditions than we know how to cure. Diagnosis can lead to serious - even lethal - forms of discrimination, especially where an unborn child is found to have an abnormal gene.

Alternatively, diagnosis can lead to treatment, including gene therapy. This booklet deals with gene therapy and other genetic interventions. The first chapter is mostly factual, with some moral comments in passing, while the second looks at the moral and religious implications of genetic interventions.

BACKGROUND SCIENCE

Genes

The body of each of us started as a single cell, when we were first conceived. Now we are made up of billions of cells, which come in very different kinds. Some are blood cells, some are brain cells, some are bone marrow cells. Most of our cells contain the same information, which is found in the *genes*; however, this information is used in different ways in different cells.

A gene is a piece of information or 'recipe' for making a protein. Genes are made of a chemical called *DNA*. DNA is stored in packages called *chromosomes*, on which the genes are located. Chromosomes come in 23 pairs, and one of each pair is inherited from each of our parents. Unlike other cells, sperm and ova (eggs) have, at some stages, one of each chromosome, rather than a pair of each kind. When the sperm and ovum come together at conception, a new human being is created who has 46 chromosomes: two of each kind. This new human being is the *zygote*, or one-cell embryo.

As the embryo grows in the body of its mother, its cells are constantly dividing. At first these cells are not committed to forming one part of the body rather than another. Cells can even separate off from the rest of the

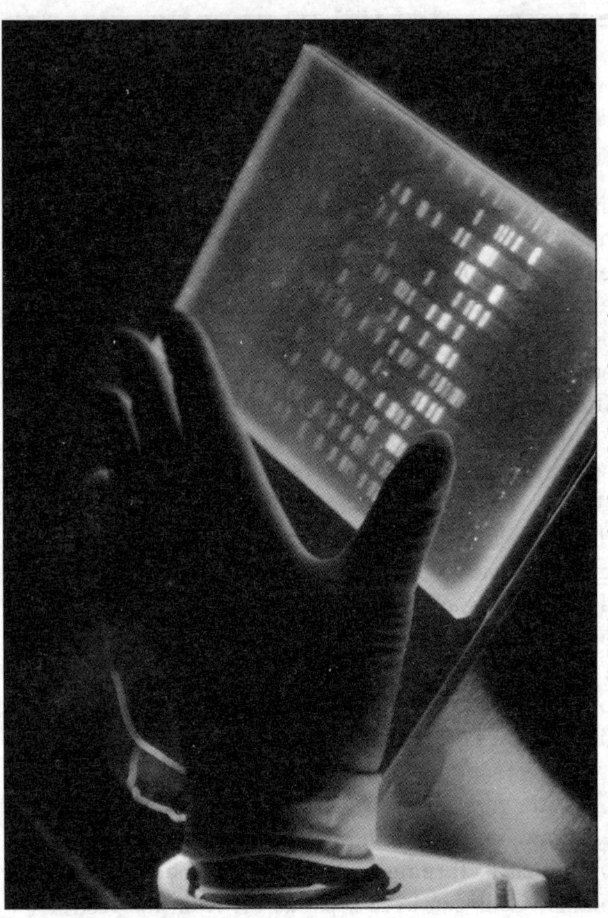

Human Genome Research: DNA sequencing by gel electrophoresis.

embryo to form a new embryo: a twin brother or sister. However, as time goes on the embryonic cells become more and more specialized into the different types of cell which the older human being needs. Some genes are switched on and others switched off, depending on what is needed by the part of the body affected. Genes work as part of the cell, and cells work as part of the body as a whole, to keep it functioning as it should.

Genetic disorders

Through the genes we inherit, the features of our parents may be passed on to us. We inherit both positive features, and also areas of weakness. For example, our sight may be bad, or we may be prone to develop heart disease in later life, as our parents may have done. For many conditions there is both a genetic component (which may involve a number of genes) and a strong environmental component. We may be much more likely to develop a disease which 'runs in the family' if we have a certain kind of lifestyle. For example, if heart disease runs in my family, I may be much more likely to develop it myself if I smoke and eat fatty foods.

Many conditions are caused by a number of genes interacting with the environment. However, some conditions, such as *cystic fibrosis*, are caused by only one defective gene. Cystic fibrosis involves a mutation in the gene responsible for forming a protein whose absence causes chest infections, and eventual damage to the lungs. Cystic fibrosis is a *recessive* condition, which means that to be affected we need to inherit copies of the faulty gene from both our parents. (Our parents may not themselves be affected, but may simply be *carriers*, so that some of their children are affected.) Other diseases such as *Huntington's* are *dominant*, which means that we would be affected even if we had only one copy of the faulty gene, inherited from one of our parents.

'Preventing' genetic disorders

Those who work in the area of genetics will often talk about *prevention* of disorders such as cystic fibrosis. Sometimes by this they mean not having children if you find you are a carrier. In principle, this could be a worthwhile option, depending on how we go about it. We could, for example, take into account our carrier status in deciding who (or if) to marry. Or a couple who are already married could choose to avoid conception through *natural family planning*, so that they do not have intercourse at times when the woman is fertile. Whether a couple have reason to avoid conceiving a child who has a high chance of having some genetic condition will depend (for example) on whether they are able to meet the child's needs.

However, what geneticists often mean by 'preventing' conditions such as cystic fibrosis is screening for the condition in the womb, and aborting any baby found to be affected. There are (as we shall see in chapter 2) very serious moral reasons against this: someone with cystic fibrosis has as much right to live as anyone else. Instead of taking the life of a child because he or she has some medical problem, we should do our best to help the child to have as good a life as he or she can.

Gene therapy

What can be done to treat those affected by cystic fibrosis? There are some treatments currently available, but none which bring about a cure. A person born with cystic fibrosis can expect to live no more than three or four decades, unless better treatments can be found. There is, however, a new way of dealing with cystic fibrosis and other inherited conditions. On this approach, we go to the root of the problem, by giving the patient a normal copy of the defective gene.

Gene therapy is a way of treating disease by delivering genes to affected cells. Scientists are working on the possibility of replacing an abnormal with a normal gene at the very same site on the chromosome. It may also be possible to build an artificial human chromosome with the normal gene already on it. This could be a way of avoiding the disruption of the work of existing DNA; it could also enable large amounts of new DNA to be delivered to the cell. So far, however, gene therapy has focused on adding the gene to the DNA on existing chromosomes, in the hope that, where it attaches, it will do good and not harm.

Germ-line gene therapy

There are two main types of gene therapy: *somatic* and *germ-line* gene therapy. Most scientists and doctors agree that germ-line gene therapy is too dangerous to consider at the present time. Germ-line gene therapy would be aimed at affecting future generations, and not just an individual patient. This would mean that, if it worked, the good effects would be inherited by the descendants of the original patient. However, it would also mean that if it did *not* work, but instead had bad effects, these bad effects would be inherited by subsequent generations.

To affect future generations, we would need to affect the *germ-line cells*; ie., sperm or ova or the cells which produce sperm and ova. For example, we could alter the ova of an adult woman, or the immature ova of a female foetus. We could also alter the less specific cells of a very young embryo: cells which would give rise to ova or sperm as well as other parts of the body.

Germ-line therapy would probably be carried out in connection with *in vitro* fertilization (creating a 'test-tube baby' in the laboratory). This would itself raise serious moral problems, as we will see at page 31. Germ-line genetic interventions could involve *cloning* or other forms of *nuclear replacement* and/or changes to *mitochondrial* genes (see pp. 14-17).

Somatic gene therapy

Somatic gene therapy involves a genetic alteration which is aimed at affecting only an individual patient. There might, in some cases, be a risk of affecting the germ-line cells, and hence the patient's descendants; however, this would not be the intention. Somatic therapy can be carried out on children, including unborn children. However, intervening on the early embryo would inevitably affect the germ-line.

Somatic therapy is much safer than germ-line therapy, and trials have been taking place for many years. Unfortunately these trials, with very few exceptions, have had little long-term success; more research is needed on how to make somatic therapy more effective. Encouraging results have been seen in some areas - for example, in helping blood vessels to grow in the legs of patients who would otherwise have needed amputations. More recently, gene therapy has been successful in treating children with very poor immune systems, who would otherwise have to live in plastic bubbles to protect them from infections.

In the future, somatic therapy may become a standard form of treatment both for *inherited* diseases and for *acquired* diseases such as cancer and AIDS. It is likely that gene therapy will be used not only to treat genetic defects such as cystic fibrosis, but also to produce gene products which are useful in the treatment of other conditions.

Genetic enhancement

Could somatic genetic interventions be used, not to treat disease, but to 'enhance' normal features - for example, height, or intelligence? Theoretically, both somatic and germ-line interventions could be used to do this. Already, the mental abilities of mice have been improved by genetic engineering. However, in practice, genetic enhancement would be far too risky in the case of human beings, even if thought to be otherwise desirable. Somatic interventions are currently proposed only for serious diseases where we need to find a more effective treatment.

Delivering genes

What are the techniques used for doing somatic gene therapy? There are various ways of delivering a gene to a cell. Viruses can be used as *vectors* (delivery systems), in view of their natural tendency to enter cells and insert their own genome into the genome of the host cell. A human gene can be inserted into the genome of a virus which has been disabled from reproducing, in the hope that the virus will then deliver that gene into the cell of the patient. Thus in the case of cystic fibrosis, a virus which targets the human airway has been used to carry the gene into the affected part of the body.

There are other methods by which genes can be delivered to cells. DNA can be directly injected into the cell; it can also be placed in fatty bubbles, which fuse with the membranes of human cells so that the DNA can enter those cells. Genes can be delivered to the patient's cells either inside his or her body, or outside the body, after which the cells are grown up and reinjected back into the patient.

Risks and benefits

Like other experimental treatments, somatic therapy is not without risk. Some risks appear to be remote, such as the risk of causing cancer or affecting the germ-line cells. Other risks, such as causing an immune reaction, appear to be higher. Until recently, somatic therapy was thought to be relatively safe; however, in 1999 an 18-year-old man died after receiving somatic therapy. Doctors will clearly need to exercise caution in doing gene therapy research.

Many genetic - as well as non-genetic - conditions could potentially be treated by somatic gene therapy. However, there are other conditions where somatic therapy does not seem appropriate. Some scientists and doctors are beginning to suggest the use of *germ-line* interventions to treat such conditions.

Mitochondrial disease

Most of our genetic material is in the *nucleus* - the central part of the cell. However, there are some genes - *mitochondrial* genes - which are situated outside the nucleus. Some people have a defect in their mitochondrial genes, which is passed on to their children. Although the nucleus is healthy, the area outside the nucleus is not.

Some have suggested that a woman who wanted to avoid passing on mitochondrial disease to her children could combine the nucleus from her ovum with the ovum from a 'donor', from which the nucleus had been removed. The 'combination' ovum would be fertilized with sperm in the laboratory. The child resulting would have two genetic mothers, or 'partial mothers' - the woman who provided the nucleus and the woman who provided the rest of the 'combination' ovum (including the mitochondrial genes).

This technique would involve *nuclear replacement*, but would not be cloning: it would not create a genetic copy of either of the women involved. A technique which is different again is already used in fertility treatment. It involves injecting mitochondria from a 'donor' ovum into the ovum of the woman who wants a child. Again, the 'combination' ovum is fertilized outside the body.

Cloning

Cloning involves the creation of a twin or copy of an individual by replacing the nucleus of an unfertilized ovum with the nucleus of a cell from that individual. The ovum is then stimulated to create an embryo. No sperm is involved, and the person created would have no genetic father. Nor would that person have a genetic mother in the normal sense of the term: the only genetic material supplied by the woman whose ovum was used (unless she cloned herself) would be the genes outside the nucleus.

In view of these mitochondrial genes, which do not come from the nucleus, the clone will not be a perfect genetic copy of the nucleus provider. Moreover, cloning itself can cause genetic changes, in addition to those genetic changes which occur naturally as the embryo matures. Finally, differences in environment between the clone and the original will lead to differences in the clone's structure and behaviour.

Cloning has already been carried out on several kinds of mammal. In 1997 came the birth of Dolly: a lamb created from a body cell taken from an adult ewe. Similar procedures have been carried out on other species. Some scientists claim to have cloned human beings; however, no human being created in this way has been transferred to the body of a woman and allowed to develop. There is a fear that, if this was done, many babies would be born disabled or would miscarry. Dolly was the sole survivor out of 277 embryos - or what seemed to be embryos - created by cloning. The great majority of mammalian clones are severely abnormal.

Cloning for birth

Cloning where the clone will or may survive to birth is sometimes called 'reproductive' cloning. The term is misleading, as there is 'reproduction' whenever an embryo is created, even if the embryo is then destroyed. A better term is 'pregnancy' cloning, or cloning 'for birth'.

Why might people want to do this kind of cloning? Perhaps so that those who carry certain genetic disorders can have unaffected children. Perhaps to allow those who are totally infertile - men with no sperm; women with no ova - to have a child to whom they are genetically related. Perhaps to allow those who want to have children without genetic input from a man (for example, single women or lesbians) to reproduce their genes.

Some might want to produce a child by cloning in order to 'replace' a dead child with a clone who would be his or her identical twin. Others might want to produce a clone of someone who had some desired feature such as high intelligence, or musical ability. Others again might want to clone themselves - perhaps thinking that they would 'live on' after death in their clone.

Cloning for research/transplantation

While cloning for birth is regarded by most scientists and doctors as too risky to consider, 'therapeutic' or 'embryo' cloning is more widely supported. Again, the term 'therapeutic' is misleading: such cloning would be far from therapeutic for the clone. 'Spare part' or 'experimental' cloning are more accurate terms. Here the aim would be to create a clone by the same method as in cloning for birth, but with no intention of transferring the clone to the body of a woman. Instead, the clone would be created solely for use in experiments, so we can harvest its cells, grow up more cells in the laboratory and find out how to use these cells to treat an older human being.

One practical problem is the lack of ova with which to create clones. Perhaps ova will become more available in the future, if tissue can be taken from a woman's ovaries and the ova in the tissue matured outside her body. Some would like to use ova from female aborted foetuses to produce clones or other embryos for research. Alternatively, some want to use animal ova, in which a human nucleus would be placed. This procedure has already been carried out in more than one country. It is not clear whether the entities created - which were never implanted in a woman's body - were real human embryos, or whether they were so abnormal as not to have human developmental potential.

Stem cell research

'Therapeutic' cloning is seen by many as a potential source of *stem cells*. Stem cells are versatile cells in the body which are able both to renew themselves and to produce more specialized cells. They can therefore be used to repair damaged human organs or tissue and perhaps, in the future, to grow up organs outside the body. There are, however, possible dangers of stem cell use in transplantation. These dangers include causing cancer, and rejection of 'foreign' stem cells by the body of the patient.

As a human being develops, stem cells become less versatile, and harder to obtain - hence the interest in 'harvesting' cells from the human embryo or foetus. Embryonic stem cells taken not from clones but from IVF embryos are already being used in research. Scientists hope that we can persuade these cells to develop into various different cell types, so that the type we need can then be transplanted to an older human being. The embryo from whom the cells are taken is killed in the 'harvesting' procedure.

Cells taken from an embryonic clone would (or could) be a close match with cells of the adult who was cloned - hence the interest in cloning for transplantation. The scientist would create an embryonic twin of the patient, who would be destroyed for the sake of its parts. Alternatively, scientists could create stem cells which would match those of the patient by combining embryonic stem cells which had been deprived of their nucleus with nuclei from the patient's cells. The result would not be a new human embryo, but cells to be used in developing transplant material. However, this procedure would require the destruction of the original embryo from whom the 'gutted' stem cells were taken. Although fewer embryos would be killed than in the case of 'spare part' cloning, since the cells of the original embryo could be used to produce many new cells, the procedure would still be dependent on the killing of some human beings.

There are other ways of getting stem cells for research and transplantation. For example, we can take cells from the patient, and transplant them back into the patient, perhaps after making genetic changes to them. We can also use stem cells from an adult donor. Already, patients have been successfully treated using adult stem cells. We have, in effect, been using adult stem cells for decades in the course of bone marrow transplants, and adult cells are now being used to treat a wide range of conditions.

Stem cells can be taken from the umbilical cord when a baby is born, and used in transplantation (these cells are also technically known as 'adult' cells). Finally, stem cells could be taken from the foetus after natural miscarriage, if it was thought that cells from a younger human donor would be more effective.

CATHOLIC TEACHING

What is the attitude of the Catholic Church to the forms of genetic engineering we have looked at so far? There have been a number of statements on genetics from popes (in particular, Pope John Paul II) and from the Congregation for the Doctrine of the Faith, a body which assists the Pope. These are not, by and large, very detailed statements; nor are they *infallible* statements (i.e. those of the most binding kind). However, they do repeat infallible teaching on respect for human life and human dignity. Moreover, the Church is - in the words of Pope Paul VI - an 'expert in humanity', whose teaching is rooted in a long-standing moral tradition, while also taking note of more recent developments. Catholics and others should therefore take seriously even the non-infallible teaching of the Church, and use it in forming their consciences.

Promoting genetic health

We will now look at statements made by the Church in relation to genetics, and at what might be the reasoning behind them. The Church is not opposed, in principle, to trying to promote genetic health. Whether this is right or wrong in specific cases will depend on how we go about it. It is not wrong, in principle, to tell those who carry genetic disorders the risk of passing on those disorders to any future child. However, it would be wrong to deprive those who carry genetic disorders of their right to get married and start a family *(Pius XII, Address to Primum Symposium Geneticae Medicae, September 7, 1953)*. It would be even more wrong to encourage couples to screen their children before they are born, and abort any child who has a certain disorder *(John Paul II, Evangelium Vitae, paragraph 63)*. The disabled have the same right to live as anyone else; they should be lovingly welcomed and supported by their families, and the rest of society.

Respecting the embryo

The Church teaches that all human beings, of whatever age, should be respected as 'persons' - i.e., as beings with full human dignity and rights. Even if we may not be sure if the very early embryo is a person, we should behave as if it were *(Evangelium Vitae, paragraph 60)*. In practice, many Church documents assume that the embryo is a person from conception. Certainly, the embryo, in Catholic teaching, should always be treated as a person: a human being with a human soul.

We should remember that the human person is not just the soul, and not just the body, but is both physical and spiritual. The soul is the body's 'form' or 'life principle' - i.e., it makes the body alive. Because the human person is not just the soul, but is also the body, we harm the person if we harm his or her living body. To quote a papal statement on genetics: "...in the body and through the body, one touches the person itself, in its concrete reality" *(John Paul II, 'The Ethics of Genetic Manipulation' Address to the World Medical Association, October 29, 1983).*

Obviously, this being the case, some genetic interventions can be ruled out at once. To produce a clone embryo so as to destroy it and use its cells in transplantation would be very wrong indeed. It would be treating the embryo as an object: as a means to an end. The same would apply to any intervention where the embryo was, or might be, deliberately destroyed - for example, germ-line therapy where the plan was to throw away those embryos for whom the treatment failed. It would certainly apply to experiments where the embryo was subjected to lethal interventions: i.e., where it was never intended to allow the embryo to survive. In the words of John Paul II:

"...to use an embryo as a pure object of analysis or experimentation is to attack the dignity of the person and the human race. Indeed, no one has the right to determine the threshold of humanity for an individual being, which would amount to claiming for himself an inordinate power over his fellow man.

"Therefore at no moment in its development can the embryo be the subject of tests that are not beneficial, or of experimentation that would inevitably lead to its destruction or mutilation or irreversibly damage it, for man's nature itself would be mocked and wounded. The genetic inheritance is a treasure that belongs or could belong to a unique being who has the right to life and integral human growth." *(John Paul II, 'Society must protect embryos', Address to a working party on the legal and clinical aspects of the Human Genome Project, November 20, 1993.)*

Gene therapy

What is the Church's position on gene therapy as such? The Church supports gene therapy in principle, including gene therapy on children. As John Paul II said in 1982:

"The research of modern biology gives hope that the transfer and mutations of genes can ameliorate the condition of those who are affected by chromosomic diseases; in this way the smallest and weakest of human beings can be cured during their intrauterine life or in the period immediately after birth". *(John Paul II, 'Biological Experimentation' Address to Participants in the Week of Study Sponsored by the Pontifical Academy of Sciences, October 23, 1982).*

The following year, John Paul II dealt with genetic interventions in more detail in a speech given to the World Medical Association. In this speech, he begins by accepting in principle a 'strictly therapeutic' intervention on the genome, providing that it 'tends to real promotion of the personal well-being of man, without harming his integrity or worsening his life conditions'.

Here the Pope does not explicitly distinguish between somatic and germ-line gene therapy. Germ-line therapy would, however, be likely in practice to pose excessive risks to those affected: a fact recognized by the Pope in a later statement (*'Genetic research must benefit every human life', Address to the Pontifical Academy for Life, February 24, 1998*).

Non-therapeutic interventions

What about genetic interventions which are not 'strictly therapeutic'? John Paul II does not condemn such interventions outright, but warns of some dangers. He starts by observing that genetic interventions must not bypass the biological and spiritual union of a couple united in marriage. In other words, the production of embryos in the laboratory so as to carry out genetic interventions on them would be morally wrong. This is an important point, which we will come back to at p.33.

John Paul II warns against laying exaggerated stress on the influence of genes on human beings. A human being is not just a body, but is also a soul; there is more to human beings than their genes. As the Pope points out in a later statement, the human being has capacities rooted in his or her nature as a spiritual being, whatever the body's deficiencies:

"The human person is not defined according to his present or future activity nor obliged to become what is glimpsed of him in the genome, but according to the essential qualities of his being, the capacities connected with his very nature. From the moment of fertilization, a new being cannot be reduced to its genetic inheritance, which are its biological basis and which hold the promise of life for the subject." *(John Paul II, 'Society must protect embryos')*

John Paul II also warns in the 1983 document against a racist mentality on the part of those involved in genetic engineering. We can think of the Nazi desire to promote the 'superior' genes of Aryans as opposed to the 'inferior' genes of non-Aryans, and of sterilization programmes - in and outside Germany - intended to prevent 'inferior' births:

"For the rest, the fundamental attitudes inspiring the intervention we refer to should not derive from a racist, materialist mentality, aimed at a human happiness which is really reductive. Man's dignity transcends his biological condition." *(John Paul II, 'The Ethics of Genetic Manipulation')*

What else might be wrong with interventions which are 'not strictly therapeutic'? John Paul II appears to be referring here not just to *preventive* medical interventions (for example, giving people increased resistance to disease) but to other, non-medical interventions - both those that *benefit* the person affected, and those that merely *change* him or her.

In the following paragraphs, John Paul II sets out some principles for evaluating 'not strictly therapeutic' interventions. Such interventions, he says, must:

"Respect the fundamental dignity of mankind and the common biological nature which lies at the basis of liberty; respect, consisting in avoidance of manipulations tending to modify the genetic store and to create groups of different people, at the risk of provoking fresh marginalizations in society.

"Genetic manipulation becomes arbitrary and unjust when it reduces life to an object, when it forgets that it has to do with a human subject, capable of intelligence and liberty, and worthy of respect, whatever its limitations may be; or when genetic manipulation treats the human subject in terms of criteria not founded on the integral reality of the human person, at the risk of doing damage to his dignity. In this case it exposes man to the caprice of others, by depriving him of his autonomy..."

"To tell the truth, the expression 'genetic manipulation' remains ambiguous and ought to become the object of genuine moral discernment, for on the one hand it covers adventurous attempts aimed at promoting I know not what superman, and on the other hand salutary efforts aimed at correcting anomalies, such as certain hereditary maladies."

There are a number of points John Paul II is making here. Why is he objecting to some, if not all, non-therapeutic interventions? Partly because such interventions would or might have bad effects in practice: e.g. in producing 'fresh marginalizations' in society. It is not hard to imagine subjects of genetic engineering who were visibly different from others, and could therefore be picked out immediately as 'genetically modified people'. Such individuals might see themselves, and be seen by others, as abnormal. We can think of a group of people who were genetically altered so as to fit them physically or mentally for doing menial tasks which other people wanted to avoid.

However, John Paul II also seems to be saying that even apart from the long-term social impact of certain genetic interventions, such interventions are morally wrong. He refers to 'modifying' the genetic store, in a way which fails to respect our 'common biological nature'. Later in the document, when speaking of genetic interventions on non-human organisms, the Pope distinguishes between interventions that 'modify' nature - i.e., that distort it - and those that 'favour its development in its own life'. And in the document of 1993, he observes that:

"Thoughtless manipulations of gametes or embryos, which consist in transforming the specific sequences of the genome that bear the traits proper to the species and the individual, make humanity run the serious risk of genetic mutations that will necessarily alter the spiritual and physical integrity not only of the human beings on which these alterations are made but even more on individuals in future generations."

Note the phrase 'proper to the species and the individual', which suggests that we are looking at changing human features to the kind of features which normal human beings (or normal human beings of that sex or race) would not have.

Imagine it was possible to give a human being some animal feature: for example, scales like a fish (this example is, of course, science fiction!) Remember, this would not be a *therapeutic* change, but one made for no medical reason. Should we not respect the kind of animal we are - the kind of nature which God has given us - and not try to make ourselves more like some other kind of animal? Can we make sense of human well-being, except as a fulfilment of our human nature? Fish have scales; human beings do not. Even if human nature is in some way limited (at least in the sense that human beings do not have scales!) is it not ungrateful - even arrogant - to refuse to accept the limitations imposed by human nature in its healthy form? It is one thing to develop ever-more-impressive tools - computers, aeroplanes, etc. - to make our own capacities more effective on the outside world. It is quite another to 'turn our tools on ourselves', and give our bodies a capacity or feature no normal human being would have.

Respect for liberty

In the 1983 document, John Paul II refers several times to *human liberty* as something we should respect in doing genetic engineering. He may be referring here to interventions which literally take away our free will: something most interventions - even harmful interventions - would not do. Like other parts of our bodies, genes can influence the way we behave, making us (for example) more aggressive. However, we normally retain our free will - at least to some extent - under emotional pressure, whether this pressure comes more from genes or more from our environment. Perhaps there is a gene which predisposes us to alcoholism, as some claim. This would not mean that people who had that gene were *programmed* to abuse alcohol - 'could not help themselves' - but merely that they would find it very much harder than others to control their drinking. An intervention on our genes which did us mental harm would probably still leave us with the capacity to choose.

However, we can imagine interventions so major as to cause severe mental handicap. In this case, the person would indeed be deprived of the exercise of his or her free will, and would be no more able to choose than a very young child. To deprive a person deliberately of his or her ability to make choices would be a serious infringement of that person's rights. Some may think that no-one would be tempted to do this; however, it has already been suggested that we could manipulate an embryo to make it develop without a brain. Such an embryo would be (supposedly) sub-human, and could therefore be used as a source of tissue for transplantation with fewer qualms than the normal embryo or foetus. Whether or not the original embryo would even survive such a drastic intervention, it would clearly be very wrong indeed to mutilate the embryo in this way.

What else could John Paul II have in mind in speaking of respect for human liberty? He appears to be thinking at least partly of interventions which *bypass* the free will of the person on whom they are performed, who is unable to give his or her consent. Such interventions may be (as the Pope says) 'capricious' - meaning, perhaps, that they do not benefit the person on whom they are performed. We can think of changing the sex of an embryo to the sex preferred by the parents: parents have no right to alter the neutral or positive features of their children in this way.

Alternatively, interventions may be 'capricious' in that they do confer some benefit, but not so as to justify the risks and lack of consent. It is one thing to intervene on a sick person who cannot give consent to restore that person's health. It is much harder to defend an intervention on a healthy person who has not consented to it, in ways which are (say) purely cosmetic. Perhaps a parent has the right, in some cases, to consent to cosmetic surgery on behalf of a child - for example, where the child's unusual appearance is making him unhappy or self-conscious. However, what about future generations, for whom we are only remotely responsible? What gives us the right to consent on their behalf? Remember that non-therapeutic changes to an embryo - or to the sperm or ovum from which an embryo is created - will affect not just that embryo, but his or her descendants down the ages.

Designer babies

It is now worth turning to another document, this time written by the Congregation for the Doctrine of the Faith, and signed by John Paul II. The *Instruction on Respect for Human Life in its Origin and on the Dignity of Procreation (Donum Vitae)* mainly deals with fertility treatments such as IVF. IVF and similar treatments are rejected as non-sexual ways of generating children which do not respect the child's dignity or that of the parents. In addition to opposing IVF, the document also opposes some types of genetic engineering or selection which could accompany IVF:

"Techniques of fertilization in vitro can open the way to other forms of biological or genetic manipulation of human embryos, such as attempts or plans for fertilization between human and animal gametes and the gestation of human embryos in the uterus of animals, or the hypothesis or project of constructing artificial uteruses for the human embryo. *These procedures are contrary to the human dignity proper to the embryo, and at the same time they are contrary to the right of every person to be conceived and to be born within marriage and from marriage. Also, attempts or hypotheses for obtaining a human being without any connection with sexuality through 'twin fission', cloning or parthenogenesis are to be considered contrary to the moral law, since they are in opposition to the dignity both of human procreation and of the conjugal union...*

"Certain attempts to influence chromosomal or genetic inheritance are not therapeutic but are aimed at producing human beings selected according to sex or other predetermined qualities. These manipulations are contrary to the personal dignity of the human being and his or her integrity and identity. Therefore, in no way can they be justified on the grounds of possible beneficial consequences for future humanity. Every person must be respected for himself: in this consists the dignity and right of every human being from his or her beginning". *(Donum Vitae, I.6)*

It is not entirely clear what is meant here by 'attempts to influence ... genetic inheritance ... aimed at producing human beings selected according to sex or other predetermined qualities.' The document appears to be referring to producing what are sometimes called 'designer babies': babies to fit our desires. This could be done either by genetic interventions on the embryo, or on the sperm and ovum which produce it, to make that embryo fit our desires, or by selecting 'high quality' sperm or ova (or sperm which will produce a child of one or other sex) so as to have a child of one kind rather than another. The document is saying that a child produced 'to order' for the parents is thereby not respected for him or herself.

There is always a danger that parents will see children as something they own and control, rather than as people with independent rights and an independent destiny. The danger is very much greater if the child comes into being in a way which inherently encourages this way of seeing children. Selection or manipulation of sperm or ova - or worse, selection of embryos - can combine with other harmful aspects of IVF to distort attitudes of parents. We can think of commercial organizations - in America, particularly - which offer sperm from Nobel-prize-winning men, or ova from good-looking women. How will the parents react if the child/product manufactured to order fails to meet their demands?

In vitro fertilization

As we have seen, the Church opposes IVF even without such further interventions as selecting or manipulating sperm or ova or embryos so as to have 'designer babies'. What is the reason for the Church's rejection of IVF *per se*? The following answer has been suggested by some Catholic writers. When a child is sexually conceived in marriage, by parents who are 'open to life' (i.e., not using contraception) the child comes into being in a way which is suited to express the parents' *unreserved self-giving*.

The couple can in this way give themselves, and accept each other, unconditionally, without 'keeping back' their fertility or their permanent commitment. Whether or not their action *does* result - or *can* result - in a new human being (they will not always be fertile) it is the kind of act which is *worthy* of resulting in a new human being. It is God who creates the human soul; however, the married couple lay themselves open to doing what may culminate, in the right conditions, in God's act of creation. The couple are prepared, by giving themselves in a way which is open to a further gift of life, to see the child as a *gift they receive*, not a *product they produce*.

In contrast, IVF embryos are not *received* by the couple as an outcome of sexual self-giving. Rather, IVF embryos are *produced*, like manufactured objects, not by an interpersonal act but by the manipulation of materials. Just as sexual conception in marriage, by an act of unconditional self-giving, prepares the couple to accept their children unconditionally, IVF has a very different structure, and tends to have a different effect. It is not surprising that IVF embryos are so often *treated* like products, when they come into being (at least on the human side) by means of a process of production. These embryos are produced in excess numbers, subjected to quality control, implanted or frozen if potentially acceptable, discarded or used in experiments if not. In short, they are treated as products or possessions of their parents or the scientists who produce them. Just as buying or selling a child would tend to make us think we owned it, the same is true of manufacturing a child.

Cloning

Cloning, if it ever became a safe and feasible procedure, would have much in common with IVF, as a form of manufacturing children involving their quality control. However, there are features of cloning which make it even worse preparation for accepting the child who would result. Apart from the features it shares with IVF, cloning also raises other problems concerning the deliberate production of a child who is a genetic copy - though not a perfect copy - of some existing person.

Children need a sense of separateness both from their parents and from others. They need to feel free to live their own lives. Sexual procreation in marriage puts the parents in the right frame of mind to respect their children's dignity and freedom. Such procreation is symbolically helpful, not only at the level of sexual behaviour, but at that of fertilization. The fusion of the parents' genetic contributions to form a new and distinct individual presents itself as at once a symbol of *relatedness*, and at the same time one of *difference*. The child is genetically related to both parents, but is still genetically unique, just as his or her life is both a new start and owes a debt to the past. The visible difference of the child from the

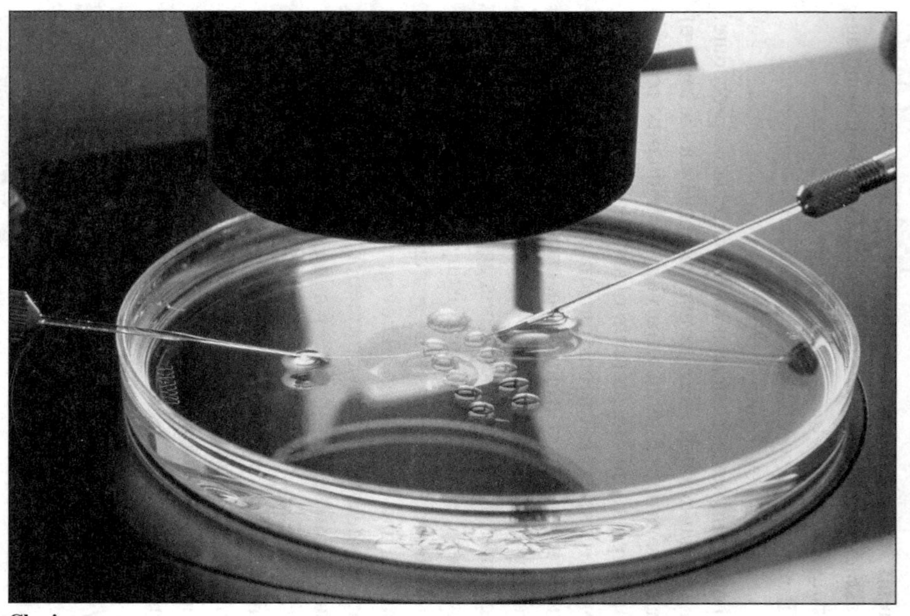

Cloning a mouse.

parents - and normally, from his or her siblings - reminds all concerned that he or she is a separate human being with a separate life to lead.

Parenthood involves - or should involve - acceptance of the child as a new human person. However, parents are all too often tempted to try to control the child in inappropriate ways, and to withhold their love or acceptance if the child is not the kind of child they want. Cloning will do nothing to help parents guard against this particular temptation, as cloning is itself a very strong form of parental control. Even in the case of natural procreation, children will often need to struggle hard to establish a separate identity from their parents. How much more of a struggle will be needed when the child is genetically identical either to one social parent, or to someone the parent or parents want reproduced?

Some may object that clones are found in nature, in the form of identical twins, and that there is nothing wrong with that. In fact, identical twins can experience emotional problems if they are brought up in ways which do not adequately recognize their separate identity. Cloning would be likely to exacerbate these problems, since whereas twins are not *deliberately created* as identical clones would be created, at least in some cases, *in order to* resemble someone else. Those who have gone to such lengths as cloning to ensure similarity between two human beings have shown, by that very fact, their wish that these two human beings be very much alike.

Family relationships

The child's sense of confusion based on his or her all-too obvious resemblance to the person cloned would be compounded by confusion with regard to family relationships. Even where two people of different sexes have been used to produce a clone, he or she will have literally no genetic parents (or other genetic relations) in the normal sense of the term. Instead, he or she will have a provider or providers of a nucleus and/or an ovum from which the nucleus has been removed. These people will not be the 'mother' or 'father' as we normally understand the terms. For this reason, they may feel even less responsibility for the child created than we find in 'donors' of sperm or ova in IVF today.

All this is assuming that the clone would be implanted in a woman's body and born as a baby. However, due to the risks of cloning to the child, which the 'Dolly' experiment has shown, it is more likely that cloning will be done, not to produce born babies, but to produce embryos so as to harvest their cells. Such a procedure is morally abhorrent, for the reasons we have looked at (see pp. 21-23). It is even worse than cloning for birth, since the clone created will not be permitted to survive. Once again, the child will first be produced, and then treated, as a manufactured object, to be used as others see fit.

Enhancement

As we have seen, cloning for birth could be used as a way of making a 'designer baby' : a child to meet specific requirements. A child who is *produced* (as in cloning and IVF) can also be produced 'to order'. In this case, the parent-child relationship is even more at risk than it is with production *per se*.

What about enhancement of children outside the context of *producing* children - i.e., enhancement of children who are sexually conceived? By 'enhancement', we mean attempts to make a human being 'better' in some way through some technical procedure. Unlike *medical* interventions, which aim to make a sick person 'better' (or to stop someone getting sick), enhancement interventions aim to make a healthy person 'better' (for example, more attractive, or more intelligent). We have already looked at the possibility of giving human beings some non-human feature, or removing some neutral or positive feature (see pp. 24-28). Here we will look again at giving human beings (i.e., those who are not sick) a higher level of some existing human trait.

Parents are generally concerned to make their children 'better' in a range of different ways. Normally they do this in ways which are not 'mechanical', but involve the child using his or her natural abilities. For example, they give the child violin lessons, or tutoring in Maths. However, parents will sometimes intervene 'mechanically' on children, intending to confer on them some non-medical benefit. Parents give children orthodontic braces to improve their appearance. They give them minor cosmetic surgery - for example, operations to pin back their ears. Some give their children drugs to improve their concentration, where these children may or may not suffer from a medical disorder.

Is enhancement by such 'mechanical' means morally wrong? It is arguably not wrong in principle, but could well be wrong in some cases. For example, the parents may accept excessive risks or burdens for the child. They may be obsessively concerned with improving the child, and unprepared to accept him or her for what he or she is (for example, if enhancement should fail). They may be unwilling to let the child develop by using his or her natural capacities - which is better, other things being equal, than developing in a more passive way, through having things 'done' to him or her. Just as health - the *possession* of human capacities - is good for human beings, so *using* our normal capacities to develop is good for human beings. Of course, in the case of many enhancements, the 'mechanical' intervention will need to be followed by the child's active use of his or her enhanced capacities. However, the initial intervention will bypass the child's own capacities, and will therefore lack the good involved in using these capacities.

The wish for enhancement - for oneself or one's children - may express a moral vice such as vanity, greed or self-absorption. It may divert people away from projects where they give each other mutual help into more solitary, depersonalized forms of self-improvement. Enhancement may be greedy if the person enhanced is already exceptionally gifted - for example, if a very intelligent person was given drugs to make him still more intelligent, or if a very good-looking person was given cosmetic surgery to make him even more so. Even if the person enhanced is somewhat *below* average in looks or intelligence, there is still a danger of enhancement encouraging the wrong preoccupations. A child may be teased by her classmates because she is somewhat slow, or somewhat unattractive: here the priority should be to change the classmates, not the child herself. Enhancement may involve a failure of respect for the person enhanced; it may also involve a misuse of time, money and attention which should have been spent on other things.

Genetic enhancement

What about *genetic* enhancement interventions - do these raise similar moral problems? In some ways, this is an academic question, as such interventions are not, at this stage, practical, at least for many features which parents (for example) might want to enhance. Features such as intelligence involve a number of genes as well as environmental factors, such that the problems enhancement would pose would be formidable. This is quite apart from the safety issues raised by such an intervention.

Genetic enhancement, like genetic therapy, could be carried out either on a single human being - a somatic intervention - or on whole generations of human beings, if we intervene on the germ-line. Leaving aside for the moment safety issues - though these issues are extremely important - genetic enhancement could well be problematic in practice, even if not in principle, just like non-genetic enhancement. Unnecessary interventions which do not, at the time they are performed, involve the child's own activity could encourage the parents to think of the child as something they control, rather than as someone who will become more and more responsible for his or her own well-being. For this reason, it is preferable for parents to attempt to influence children where possible at times when they are able to choose or, at least, to react, rather than (as with some genetic interventions) before they are even conscious. Parents need to be encouraged to take a flexible view of the child's future, to be interactive rather than manipulative, to focus on the child's real needs, rather than the parents' own desires. And if genetic enhancement took place in the context of 'producing' a child, it would involve all the problems with 'designer babies', and rejection of 'substandard' babies, which we looked at earlier.

Some have suggested that genetic interventions could be used to improve a person's *moral* performance: for example, to make him or her less aggressive. However, we should be cautious in this area. While it may theoretically be possible to remove 'mechanically' certain temptations (e.g., to aggressive behaviour), we should remember that good moral character will only be developed by our subsequent choices. Moreover, the choice to resist temptation which the 'non-enhanced' person can make is itself morally valuable, strengthening the virtues he or she is showing under stress.

Germ-line genetic enhancement

What about genetic enhancement intended to affect the germ-line (and hence future generations)? There are special reasons why enhancement affecting the germ-line would be morally wrong. We do not have the same responsibility for strangers (i.e., future generations) as we do for ourselves and our children. Just as it would be intrusive for the State to take children off the street and put braces on their teeth, it would be similarly intrusive (and in practice far too dangerous) to attempt a non-therapeutic change to the germ-line. It is one thing to intervene on strangers who cannot consent in a medical emergency. It is quite another to intervene on strangers - in this case, future generations - where there is no pressing medical need.

Germ-line therapy: moral concerns

This brings us again to the subject of therapeutic (i.e., medical) interventions. As we saw earlier, germ-line therapy is generally assumed to be too dangerous in our current state of knowledge. There are dangers both for the immediate subject - the embryo on whom we intervene, or who results from 'modified' sperm or ova - and for his or her descendants. It is true that doctors are sometimes entitled to take high risks in treating patients where there is no other chance of curing an otherwise fatal disease. However, in the case of germ-line therapy, those we are treating might very well not have existed had such therapy not been available. The parents of the child who is immediately affected might only have conceived a child at all because germ-line therapy was on offer. If germ-line therapy had not been on offer, they might not have chosen to risk having children with the relevant condition. More risks are justified in the case of treating an existing sick child or adult than would be justified in the case of a child who would only be conceived if a certain treatment was available. While the decision not to have a child is one which may involve considerable sadness, it would not be acceptable to prevent such sadness in couples by offering therapy which carried undue risks. Another factor to consider would be the cost of developing germ-line therapy; this, too, would be harder to justify where it was not a case of treating a person who would in any case exist.

However, even if the cost and the risks of germ-line therapy could be brought down to acceptable levels, parents who chose germ-line therapy for their children might have other questions to consider. For example, imagine germ-line therapy had become a very safe procedure, but had become so only by means of destructive experimentation on human embryos. If a couple accepted germ-line therapy, might this not give other people the impression that they had no problem with the embryo experiments by means of which the therapy was developed? The risk of giving this impression, thereby setting a bad example, would be something the couple would need to consider in deciding whether to accept the treatment. It is also worth remembering that there could be another option available in the future, which would help some couples to have children unaffected by genetic disorders: the screening of unfertilized ova, following which normal ova would be replaced in the woman's body in the hope that a child would then be naturally conceived. Such a procedure could, in the future, be a safer and morally preferable alternative to germ-line gene therapy.

What are some other moral problems germ-line therapy could raise? As we have seen, germ-line therapy might well take place in the context of IVF, which is itself morally wrong. Moreover, those embryos for whom the treatment failed would be likely to meet the fate of other 'abnormal' IVF embryos, in that they would not be transferred to the mother but thrown away or used in research.

Germ-line therapy could also raise problems concerning genetic relationships. Earlier (at p. 14), we looked at the suggestion that women with mitochondrial disease could have IVF using only the nucleus from their own ovum, which would be placed in the ovum of another woman prior to fertilization. Such a fragmentation of motherhood would be morally wrong, quite apart from the issue of the non-sexual production of the child. We should not deliberately create a child with more than one candidate for the role of mother; nor should women help conceive children they do not intend to bring up. Children need to feel secure in their identity, and accepted by their parents. 'Rival' parents who take no part in their upbringing will do nothing to provide them with this sense of security.

Moreover, similar problems of 'rival' parents could arise even if the child was sexually conceived. Take the following procedure, which has been performed on laboratory animals, and could in theory be performed on men who were carriers for genetic disorders. The procedure would not be germ-line therapy as normally understood, though it would be aimed at preventing the transmission of genetic disease. Instead of intervening on *some* of a person's genes, *all* that person's reproductive cells, with all the genes they contain, would be destroyed. Spermatogonia (the cells which produce sperm) would be taken from the testes of a male 'donor' and transplanted into the testes of the man whose own spermatogonia had been destroyed. The man receiving the transplant would then conceive children by sexual means, but the children he conceived would be the genetic children of the man from whom the transplant came. In having the transplant, the patient would not intend to give himself entirely to his wife in having intercourse, but to 'hold back' his fertility, substituting the fertility of another man. The child would have two rival fathers, which is harmful to the child's sense of identity, as we know from the experience of children conceived with the use of 'donor' sperm. Moreover, the 'donor' of the spermatagonia would be acting - consciously or otherwise - as an irresponsible parent, in that he would deliberately help conceive children he would never see or help to bring up.

Somatic therapy: moral concerns

Finally, to return to somatic therapy: there are moral concerns even here. Such therapy does pose certain risks, as we saw earlier (p. 11). In view of these risks, it seems right to restrict somatic therapy to cases of serious diseases where there is no satisfactory alternative treatment. It should be remembered that somatic therapy is still at a very early stage. The patient needs to be given a realistic picture both of risks and of likelihood of benefit, so as to get informed consent from him or her without raising false hopes. Those involved in trials of somatic therapy should not be required (as they sometimes are) to use contraception to reduce any risk to the germ-line. If there really is such a risk, it can be reduced by using natural family planning. As far as costs are concerned, somatic therapy needs to be justified in the context of other claims on healthcare resources. And if somatic therapy is tried first on animals - which would help reduce the risks to human beings - the animals must be treated humanely.

Conclusion

To sum up: what should be the attitude of Catholics to human genetic engineering? The genome is a part of the body which may, in principle, be healed, like other parts of the body. However, as a part which can affect many other people's bodies, it needs to be treated with special care. Both somatic and germ-line therapy raise moral problems, but in the case of germ-line therapy these problems are currently insurmountable. The problems include the manufacture of a child by such means as IVF, the likelihood that 'failed' embryos would be destroyed after therapy, the need for prior embryo experimentation and the risks to the subject and to future generations. These risks would be harder to justify in view of an alternative to germ-line therapy, and to passing on a genetic disorder: the choice to avoid having children.

In our search for treatments for genetic disorders, we should keep a sense of perspective. Genetic disorders are bad in themselves, and it is good to look for a cure. However, illness is not the ultimate evil, and good can come from illness both for the person affected and for the rest of society. There is nothing shameful in being dependent on others, as many people inevitably are; on the contrary, this can build community between human beings. While the Church, in particular, has always been involved in curing the sick, she has also been involved in supporting those who cannot be cured. The Church has also emphasized what many sick and disabled people themselves believe: that much good can come from accepting our bodily limitations.

Christians and others should be careful to ensure that in trying to treat or prevent some condition, we continue to value those human beings who have that condition. All too often, in clinical genetics, the affected individual is treated as something to be literally discarded if he or she cannot be cured. All human beings must be respected, whether they are adults or children, including children in the womb or very young human embryos. Christians are called to love justice; to show special concern for the weak and defenceless; to be alert to what meaning there may be in human suffering, even while trying to overcome suffering which can be avoided. Concern for genetic health must not deflect attention from our need to turn to God: spiritual health will always be more important than genetic health. All the more reason, then, to promote genetic health only in ways which respect human dignity, and are in accordance with God's plan.

Further reading

John Paul II, *'The Ethics of Genetic Manipulation'* (Address to the World Medical Association, October 29, 1983)

John Paul II, *'Society must protect embryos'* (Address to a working party on the legal and clinical aspects of the Human Genome Project, November 20, 1993)

Congregation for the Doctrine of the Faith, *Donum Vitae (Instruction on Respect for Human Life in its Origin and on the Dignity of Procreation)*

John Paul II, *'Genetic research must benefit every human life'* (Address to the Pontifical Academy for Life, February 24, 1998)

Genetic Intervention on Human Subjects: The Report of a Working Party (Catholic Bishops' Joint Committee on Bioethical Issues, 1996)

GLOSSARY

Abortion: The deliberate killing of an unborn child.

Ameliorate: To improve.

Anomaly: Irregularity; abnormality.

Carriers: People who are not themselves affected by a genetic condition, but may pass it on to their children.

Cell: A basic unit in the structure of living things. Some organisms are made up of only one cell, whereas others are made of many cells, at least at some stages in their lives.

Chromosomes: Bundles of DNA.

Cloning: The production of a genetic copy of an existing human being (or other organism). Cloning is often called 'reproductive' where the aim is to bring the clone to birth, and 'therapeutic' where the aim is to use the clone at the embryo stage for research or transplantation.

Conception: The beginning of a new human life; fertilization.

Condition: Illness; disorder.

Congregation for the Doctrine of the Faith: The Vatican body which is concerned with safeguarding and developing Church teaching.

Conjugal: Concerned with marriage.

Contraception: Something used to prevent sexual intercourse from leading to the conception of a child.

Cystic Fibrosis: A genetic condition characterized by abnormally thick mucus, leading to chest infections and eventual damage to the lungs.

Diagnose: To identify a disease.

DNA (deoxyribonucleic acid): The chemical of which genes are made.

Dominant: A genetic condition is dominant where an abnormal gene inherited from one parent is sufficient for the child to be affected.

Donum Vitae: Document by the Congregation for the Doctrine of the Faith on IVF and similar forms of fertility treatment.

Embryo: A human being from its origin (normally fertilization) until it is 8 weeks old.

Enhancement: The attempt to improve a human being by some technical, non-medical intervention.

Fertility: In humans, the ability to conceive/give birth to children.

Fertilization: The union of (male) sperm and (female) ovum to generate a new human being.

Foetus: An unborn human being between the age of eight weeks and birth.

Gametes: The reproductive cells: sperm in men; ova in women.

Gene: The basic unit of heredity by which traits are passed on from one generation to the next.

Genetics: The study of genes and their effects on living things.

Genome: An individual's genetic makeup.

Germ-line cells: Sperm or ova or the cells which produce sperm or ova (including the cells of the early embryo).

Gestation: Pregnancy.

Human Genome Project: Co-operative work by scientists on the location of genes on chromosomes and the definition of their chemical structures and their function.

Huntington's Chorea: A genetic condition in which, during adult life, there is worsening involuntary movement and progressive dementia, and which leads eventually to death.

Immune reaction: The body's rejection of (for example) cells from some other organism.

Infallibility: The authority given by God to the Pope and Bishops to teach without error on faith and morals in certain situations.

Intrauterine: Within the womb.

IVF (*in vitro* fertilization): The fertilization of an ovum outside a woman's body to create an embryo. An IVF child is sometimes called a 'test-tube baby'.

Malady: Illness.

Membrane: Tissue surrounding a cell.

Menstrual cycle: The monthly cycle of fertility in women.

Mitochondrial genes: Genes outside the cell nucleus.

Mutation: A change which occurs in the genetic material of a cell, which may cause that cell and cells derived from it to differ from those of the normal type.

Natural family planning (NFP): A form of family planning acceptable to the Catholic Church. It involves the couple choosing to have or avoid having intercourse at times in the woman's menstrual cycle when she is fertile.

Nucleus (plural nuclei): The central part of the cell, containing most of the cell's DNA.

Organism: A living being; a self-organizing whole.

Ovary: The organ which produces ova.

Ovum (plural ova): The egg or female reproductive cell, which may be fertilized to create an embryo.

Parthenogenesis: The production of an embryo from an ovum, with no contribution from a sperm.

Person (human): Someone who possesses full human moral status.

Protein: An essential ingredient in all living things, which the genes help to produce.

Recessive: A condition is recessive when a gene exercises little or no outward effect unless it is present in both chromosomes; i.e., inherited from both parents.

Right to life: The right of every innocent human being not to be killed deliberately, or to have his or her life unfairly endangered.

Screening: Testing for a disease.

Somatic cell: Any cell of the body except a germ-line cell. Changes to somatic cells affect only the individual whose cells they are.

Soul: The life-principle of human beings, which survives death.

Sperm: The male reproductive cell, which may fertilize an ovum to create an embryo.

Spermatogonia: Cells which produce sperm.

Stem cells: Versatile cells in the body which are able both to renew themselves and to produce more specialized cells.

Testis (plural testes): Testicle; the organ which produces sperm.

Trait: Feature; characteristic.

Twin fission: The splitting of an embryo to produce identical twins.

Uterus: The womb.

Vector: A delivery system for inserting a gene into a cell.

Virus: A very small parasite which can reproduce itself within a living cell.

Zygote: An embryo at the one-cell stage.

CTS
MEMBERSHIP

We hope you have enjoyed reading this booklet. If you would like to read more of our booklets or find out more about CTS - why not do one of the following?

1. Join our Readers CLUB.
We will send you a copy of every new booklet we publish, through the post to your address. You'll get 20% off the price too.

2. Support our work and Mission.
Become a CTS Member. Every penny you give will help spread the faith throughout the world. What's more, you'll be entitled to special offers exclusive to CTS Members.

3. Ask for our Information Pack.
Become part of the CTS Parish Network by selling CTS publications in your own parish.

Call us now on 020 7640 0042 or return this form to us at CTS, 40-46 Harleyford Road, London SE11 5AY
Fax: 020 7640 0046 email: info@cts-online.org.uk

❏ I would like to join the *CTS Readers Club*

❏ Please send me details of how to join CTS as a *Member*

❏ Please send me a *CTS Information Pack*

Name:..

Address:...

....

Post Code:...

Phone: ....

email address: ....